Discovering
WINDMILLS

John Vince

Line drawings by the author.

Shire Publications Ltd.

1977

CONTENTS

During the past eight years a number of mills have attracted the attentions of restorers. Among the outstanding renovations are Friston (Suffolk), Lacey Green (Bucks), Stansted Mountfitchet (Essex) and Wilton (Wilts). To all those involved in these and other windmill projects posterity will owe a great debt. The best immediate tribute to their work is a constant flow of visitors to the mills they have laboured to save.

The chapter on notable windmills has been extended, but the author will be pleased to know of any other mills open to the public which have not been included. Similarly, the Gazetteer has been revised once more but further additions and corrections will be gratefully received.

The author is grateful to all those readers who have written to him about previous editions of this book. In preparing the fourth edition he is particularly indebted to: J. Cooke, Peter Dolman, Cedric Greenwood, Kenneth E. Hartley, Piers Hartley, J. Hughes, V. H. F. Knight, Andrew Sillence and Arthur Smith.

Photographs are acknowledged as follows: Jeffery W. Whitelaw, plates 1, 7, 8, 20; Cadbury Lamb, plates 2, 3, 6, 8, 11, 13, 14, 15, 16, 17; John Vince, plates 4, 5; Victor Knight, plates 9, 10, 12, 18, 19.

INTRODUCTION

Few things add as much atmosphere to the English countryside as a windmill. Even in decay a mill possesses a dignity few other buildings can equal. The history of windmills in England stretches back to the turbulent days of the 12th century. One tradition suggests that they were introduced into this country by Crusaders returning from the wars. This may be partly true, but all that we can say with certainty is that windmills were first built here some eight centuries ago. None of the original structures remains of course; but among the mills which have survived to the present are some that are at least 300 years old and many more over a century.

In feudal times villagers were compelled to take their hard won grain to the Lord of the Manor's mill to be ground. The miller too took a toll for his work, and it is easy to see how the traditional mistrust of the unfortunate miller came about. As Chaucer observed—

"His was a master hand at stealing grain.
He felt it with his thumb and thus he knew
Its quality and took three times his due. . . ."

The number of windmills still at work today is a minute fraction of the many hundreds which once helped to grind the nation's grain. When steam power was applied to flour milling in the middle of the last century the decline of the windmill began. Steam power could be generated at will and had many advantages over the wayward wind. As a result, millers introduced steam plants to supplement their mills' grinding capacity, and when a mill itself reached the point where it required expensive repairs the steam mill supplanted it.

The final blow to the windmillers' craft came when steel-roller mills began to dominate the flour trade. From then onwards the miller was reduced to grist milling alone, and it is not surprising to find that many mills were finally forced out of business just after the 1914-18 war.

During the 1930s time and the elements destroyed mill after mill and an important part of our national heritage slipped almost unnoticed into oblivion. The most vulnerable structures were the wooden post and smock mills, but the brick and stone tower mills also disappeared when the materials they contained were needed for re-use elsewhere.

Windmills represent an important part of our technological history; and while it may not be possible or desirable to try to preserve every mill there are many sound reasons in favour of ensuring the survival of some. Posterity is entitled to a share of the heritage we may have either ignored or enjoyed.

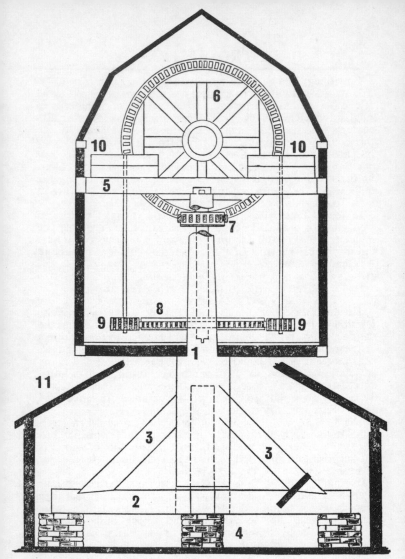

Post Mill Diagram. *Typical Hurst Frame arrangement with underdriven stones:*
1. Main Post. 2. Cross Trees. 3. Quarter Bars. 4. Brick Piers. 5. Crown-tree.
6. Brake Wheel. 7. Wallower. 8. Great Spur Wheel. 9. Stone Nut. 10. Stone.
11. Roundhouse.

POST MILLS

Although the earliest mills have long since disappeared we can say with some degree of assurance what they looked like. Representations of windmills have survived in manuscript records, carvings, stained glass, plasterwork and paintings.

One of the oldest windmill pictures is an engraving (*circa* 1349) on a memorial brass in the church of St. Margaret's, King's Lynn, Norfolk. Good carvings in wood (*c.* 15th century) can be seen on a misericord at Bristol Cathedral, and on a 16th century bench end in the parish church of Bishop's Lydeard, Somerset. A fine 15th century windmill picture also survives among the stained glass of Fairford church, Glos. Each of these illustrations shows the simplest type of windmill —the post mill.

The wooden framed body of the post mill was supported by a massive upright post. This was held in position by quarter bars which rested upon horizontal cross trees. Most post mills had two cross trees at right angles to each other; and the ends of these were usually supported by short pillars of masonry which allowed the air to circulate freely and keep the timbers dry. As the diagram shows, the post did not rest upon the cross trees but was suspended a little above them.

There were, however, at least five post mills which had three cross trees instead of the normal two, and six quarter bars in place of the usual four. The last mill of this kind stood, until late 1967, at Chinnor, Oxon. Its tottering frame was dismantled to make way for a house. The other mills which had the same type of trestle were at Bledlow Ridge (demolished *c.* 1933) and Stokenchurch (collapsed 1926), both in Bucks; Moreton, Essex (demolished 1965) and Costock, Notts.

A mill which retains the simple lines of the early post mills can be seen at Bourn, Cambs. It has plain gabled ends and an open trestle. The precise age of this mill is not known but it must rank among the oldest surviving specimens.

The mill which can, at present, claim to be the oldest dated mill in England is at Pitstone, Bucks. One of its timbers bears the date 1627. The mill was given to the National Trust in 1937 by its former owner Mr. L. J. Hawkins. Its working life ceased when it was badly damaged in a severe storm in 1902. During the past few years this splendid mill has attracted the attention of a group of local enthusiasts who embarked upon an ambitious scheme to restore the mill to its former grandeur. After a sustained effort this must rank among the best kept mills in England. Yet not many years ago it had been in danger of extinction.

In common with most timber framed buildings the members of a post mill usually have the principal joints numbered. This made erection an easier operation especially if the timbers were prepared in one place and transported to another for assembly. Roman numerals were often used to mark joints, but other symbols, based on the ancient Runic characters, were also employed by some craftsmen.

Although some post mills still have an open trestle most have had a brick roundhouse added to provide extra protection against the elements. The addition of a roundhouse provided more storage space; and it also encouraged certain writers to refer to mills with this feature as turret post mills.

One of the miller's daily tasks was to adjust the mill's position so that the sails faced the wind. To make this task easier post mills have a long tiller beam which projects from below the superstructure at the rear. An added refinement was a cart wheel at the end of the beam. When mills of this type were in use a series of posts were arranged around the circumference of the circle described by the tiller beam to provide fixed anchor points. Another method of revolving the carefully balanced mill was to harness a horse to the tailpole.

With the progress of time a mechanical and automatic method was introduced to enable the mill to adjust itself to any change in wind direction. The device was a fantail and for post mills there were three different methods of attaching them. A few mills—as at Nether Dean, Beds—had fantail gear attached to the mill body above the rear gable (this mill was really a hybrid). Such a method does not seem to have been the most popular probably because the driving mechanism was not readily reached for repair and maintenance. The second method employed a separate framework or carriage—as at Cross-in-Hand, Sussex—which was fixed to the tailpole A fan carriage usually had two iron wheels at ground level. The third way was to erect a suitable framework upon or above the mill ladder. The two latter methods were the most popular, and they both required a rail track or paved circular path around the mill to assist their passage. It is interesting to note that the automatic fantail was not extensively adopted in Europe.

Another device sometimes fitted to the tiller bar was a simple lever—a talthur. A short chain was attached to this and to the lower part of the ladder. When the mill had to be luffed, to move the sails into the wind, this lever enabled the miller to raise the steps clear of the ground.

Many post mills were extended during their working careers, but they were designed in such a way that the only possible form of extension had to be on the rear gable forming a simple lean-to. Among those with this sort of extension were Argos Hill, Sussex and Great Chishill, Cambs.

Post mills originally had common sails (see the later chapter on sails), but in time a large number of them were modernised by the substitution of patent or spring set sweeps.

Their original gearing was entirely wooden, and when it later required replacement cast iron cogs were frequently used.

In many ways the post mill is perhaps the most picturesque of our English windmills. Even when they begin to totter their grace and dignity is not lost. Forgotten and isolated, a number of mills struggle for their existence, but year by year they slowly bend to the will of the wind. Too soon they will be gone, and men will forget where they once stood.

Other mills have been more fortunate, and for those which have attracted the attention of private or public benefactors we should be grateful. A mill restoration usually depends upon local dedicated enthusiasts, and without the efforts of many such groups the English countryside would be very much the poorer. Few things can complete a prospect so well as a wind-mill—and where proportion and elegance are concerned other mills, for all their charm, cannot equal the lines of a post mill.

One of the taller post mills under repair at the present time is at Framsden, Suffolk. This fine mill, with its ogee gable, dominates the village skyline. Few can fail to admire it as it soars above the cottage pantiles. East Anglia is a Mecca for the windmill hunter, and one of its treasures is undoubtedly the mill at Saxtead Green which is open to the public. The mills at Friston and Thorpeness are also excellent examples and follow the same Suffolk tradition.

The splendour of these working mills impressed William Cobbett as long ago as 1830. Writing about his approach to Ipswich he noted that "the windmills on the hills . . . are so numerous that I counted, whilst standing in one place, no less than seventeen. They are all painted or washed white; the sails are black; it was a fine morning, the wind was brisk, and their twirling altogether added greatly to the beauty of the scene . . . and . . . appeared to me the most beautiful sight of the kind that I ever beheld" (Rural Rides).

Not all mills were painted in the manner Cobbett describes; in Buckinghamshire, for example, there seems to have been only two like this. One of these stood at Great Missenden and was burnt down in 1876. Fortunately the artist E. J. Niemann recorded it in a fine landscape he painted in 1868.

7

SMOCK MILLS

The alternative to turning the entire mill body to face the wind was to provide the mill with a revolving cap to which the sails were fixed. History has neglected to record the name of the genius who invented the method of attaching the sails and windshaft to a mobile unit that was both lighter and easier to manipulate, but his innovation allowed mill design to evolve a stage further.

The development of the smock mill represents the second stage in windmill design. Smock mills, so we are told, derive their name from an alleged resemblance to a man in a smock. Whatever the truth of the matter the name has stuck, with the variation frock, and wooden framed tower mills retain it to this day.

Post mills are constructed on a rectangular unit, but the smock mill design is derived from a circle. Wooden members, which had to be straight, reduced the circle to a series of connected chords. The usual pattern of smock mill construction was based upon eight sides, but six, ten and twelve sided mills have also been constructed. This departure from square jointed construction must have produced problems for the millwrights to solve. The key structural members of a smock frame are the stout cant posts which form each corner. It must be remembered that these posts are closer together at the top than they are at the base. To form good joints that were leaning in two dimensions must have called for all the carpenters' skill. To prevent the base of the cant posts from rotting too quickly the smock towers were often raised on a brick plinth. This had the effect of prolonging the life of the posts, but in time many mills ran into serious difficulty when a corner post became weak and threatened the stability of the whole structure.

The diagram opposite shows the main structural members of a typical smock mill. One of the essential needs of a smock and a tower mill is a level and stable curb upon which the cap may rotate freely. From the way the cant posts are arranged they naturally tend to want to move outwards and allow the structure to collapse. Even small movements within the frame could allow the curb to buckle out of true, and anything which hindered the free movement of the cap was an anxiety to the miller. Perhaps the greatest hazard arose when a cap jammed and the wind direction changed suddenly. If the wind was very strong and blew from behind the sails, making the mill tail-winded, the consequence could be very serious

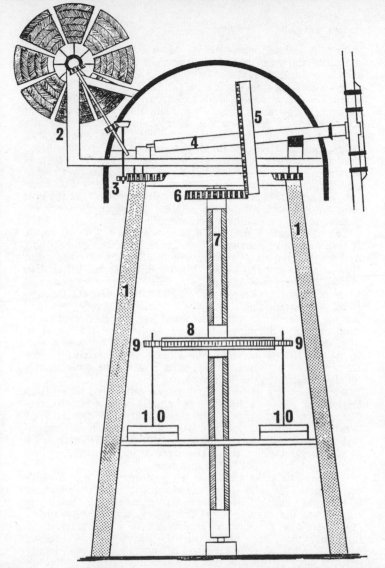

Smock & Tower Mill Diagram: *1. Tower wall. 2. Fantail and gearing. 3. Drive to curb. 4. Wind Shaft. 5. Brake Wheel. 6. Wallower. 7. Main Shaft. 8. Great Spur Wheel. 9. Stone Nut. 10. Mill Stone.*

indeed. A tail wind could blow down a post mill, or remove a smock mill's cap and sweeps or sails in far less time than they would take to replace.

In some ways smock mills are far more vulnerable than post mills. They have sloping sides and usually twice the number of corners, if not more; these all provide additional chances for the elements to penetrate and destroy. Joints in the weatherboarded cladding were sometimes sealed with zinc or lead strips, but in spite of these precautions the miller always needed to keep a weather eye on his mill. The provision of windows in the sloping sides of a smock mill also gave rise to problems, and where the window frame joined the body special care was needed to make it weatherproof. Like the post mills some smock mills were painted, and were therefore expensive to maintain; others were tarred.

The oldest smock mill in England is at Lacey Green, Bucks. It was originally constructed at Chesham round about 1650, and was moved to its present site in 1821. Most mills suffer from a move, but the Lacey Green mill has lasted better than many built long after. Its recent condition was very poor and a complete restoration is now being carried out by the Chiltern Society. The machinery is almost certainly the oldest wooden gearing in the land, and its size and age together must make it of unusual archaeological interest. Quite apart from its technical attributes this mill commands one of the finest prospects in the Chiltern countryside; and when it is protected from the elements and refurnished with sails it will add more than a touch of dignity to the landscape.

We cannot say when the first smock mill was built in England, and it seems as if they were actually invented in Holland. The oldest dated smock mill is probably the one at West Wratting, Cambs.—1726. The last mill to be built in England was a smock mill. It was raised in 1928-29 by Holman Bros. at St. Margaret's Bay, Kent. The same county boasts the tallest mill in the land—High Mill, Cranbrook.

Some smock mills employed a counterpart of a post mill's tiller beam. Tail poles were a common feature on many smock and tower mills. They were strengthened by outriggers attached to the cap. To assist the luffing process some smock mills were also fitted with fantails. These were lighter structures than those required to turn the heavy post mill bodies. Mills often had a second method of rotating the cap—usually effected with an endless chain, which operated a pulley on the fantail mechanism, or with internal gears.

Early smock mills were probably, like post mills, originally fitted with common sails. At a later stage spring or patent sails were substituted on many mills, but a few smock mills demonstrated the innate conservatism of certain millers who used both a pair of common sails and a pair of spring or patent sweeps—as at Earnley, Sussex and Lacey Green, Bucks.

If a mill was made with a tall tower—perhaps to avoid the effects of other buildings close by—the sails might be too high to reach from the ground. So a new feature arose in the design, and a tall mill had a gallery constructed around its sides to allow the miller to attend to his canvas, as at Cranbrook.

Smock mills often had a tarred base, and they always seem particularly attractive if their superstructure is painted white. A number of smock mills have survived to adorn the countryside, and some of them serve as homes.

A striking example is Ibstone, Bucks. In 1967 a mock fantail and sails were temporarily added to it for effect by a film company which used the mill for some location shots in the film "Chitty Chitty Bang Bang". Whatever the purists may say about mock sails there can be no doubt that the mill looked all the better for them. It finished work in 1914-18, and its present condition suggests that it will last for many years to grace the skyline above the picture-book village of Turville. The mill was formerly owned by St. John's College, Oxford.

TOWER MILLS

The difference between a smock mill and a tower mill depends upon the materials used. A smock mill has a wooden frame, and a tower mill is usually constructed from stone or brick. Both have the same basic shape and are arranged internally on the same principles. The slope of the mill wall is called the batter. Most towers are round in plan, but it is not unusual to find octagonal towers, as at Wheatley, Oxford. Another variation was to make the lower part of the tower octagonal in plan and the upper portion round, as at Great Bardfield, Essex.

Tower mills represent the pinnacle of mill design. They were probably introduced at about the time of the first smock mills and, generally speaking, are of larger proportions than other types. Various materials have been used to protect them, and many stone towers were enhanced by a veneer of tiles or slates. The wall of the tower mill was usually very thick and, like some of the smock mills mentioned above, a good num-

11

ber of them have been converted into homes. The octagonal brick tower at Wendover, Bucks, became a residence in 1940; and its present owner still lives in comfort under the largest mill cap in England. Stone or brick towers were easier to weatherproof than the boarded smock mills. The owner could employ a cement render if the brickwork began to crumble. This was not always visually pleasing, but it answered the miller's purpose. One other safeguard was to paint the tower with tar. A good coat could last for many years and it was a sound practical answer to the driving rains of an English winter. Some of the most attractive towers are those with tarred sides and white paintwork, as at Heckington, Lincolnshire.

Tower mills had many advantages over post and smock mills. One poor feature however arose in towers which had windows and door openings placed one above the other. This arrangement allowed lines of weakness to develop if the foundations subsided even a little. A superior and safer method was to place openings around the body in a spiral fashion. Almost any tower mill with vertically spaced openings shows signs of some subsidence in its structure. A dropped window arch could easily lead to the distortion of the curb and this all too often made it impossible to revolve the cap, an expensive thing to correct. In some circumstances a jammed cap could mean the final closure of the mill.

Like smock mills the tower mills were provided with common, spring or patent sails according to their age and prosperity. Most of them had a gallery and even those that used common sails until the end of their days usually had a fantail mechanism. The finest tower mills were built at the end of the 18th century and in the early part of the 19th. This was the great age of tower building, but even late in the last century some millers were wealthy enough to restore, refit or rebuild their mills.

In many ways tower mills are more durable than their wooden counterparts, but they can disappear without trace—like the flint rubble tower at Holmer Green, Bucks, which collapsed in 1929 after a century of life. This mill was unusual in this part of the country—it had a triangular gabled cap which was turned by an endless chain on a pulley; after the style of Melin-y-Bont, Gwynedd and mills on the Fylde, Lancs. Mills built of coursed stone or brick stand more chance of survival when they fall into disuse. A great number of long redundant towers still almost intact add interest to the English countryside. Others for one reason or another survive in a truncated form. The remains of Great Horwood mill, Bucks, stand, inhabited by pigs, behind a modern barn. The super-structure was dismantled at the beginning of the last war when an airfield was built close by and the tower was considered to be a hazard to aircraft. A few miles away the stump of Stewkley mill also survives; it was taken down to its present level in 1922 when its upper part was in a dangerous condition.

A study of Ordnance Survey maps suggests that when a mill is partly dismantled it is then, quite properly, omitted from the next revision of any given map sheet. The seventh edition of the O.S. One Inch series employs two symbols for wind-mills—one indicating mills 'in use' and the other for 'disused' mills. Partial remains are not shown and the diligent searcher will need to use an older edition as well as a new one if he is to get the best results from his wanderings. Post and smock mills do not often leave many visible traces behind when they finally fall to the ground, but the base of a tower mill may quite often survive behind a hedgerow or the miller's house.

All three types of mill have been erected on man-made mounds which were constructed to add effective height to a mill structure. On crowded sites where, perhaps, large barns or other buildings could affect the free passage of the wind such an arrangement was a satisfactory and less expensive alter-native to the erection of a high structure. Where a mill has completely vanished a mound will often remain as it is a costly item to eradicate. In remote places known mill mounds have been slowly, over many years, destroyed by the plough. A good specimen of a mill mound with its tower may still be seen at Blackthorn, Oxon.

DRAINAGE MILLS

The work of draining the Fens began as far back as the 16th century. Wind power was set to work to aid man's efforts to reclaim valuable areas of land and from this need the drain-age mill evolved.

Earlier drainage mills were plain smock structures, but later specimens were built of brick and were often tarred. Drainage mills usually had a Dutch look about them with their dumpy towers, common sails and tail poles. In fact, the design seems to have originated in Holland. Latterly patent sails and fans replaced the tailpole.

Internally their machinery was simple. The wind turned the sails and the windshaft. A wallower, fixed to the vertical shaft, was in turn driven by the brake wheel, and so the shaft was rotated. At ground level a bevel gear was connected to the pit wheel, which was mounted on the same axle as the paddle wheel. The revolving paddle wheel scooped up water which it discharged at a higher level.

Marshmen lived with their families in these often remote mills. Many were so isolated that the only way to reach them was by water. The decline of the drainage mill is closely linked with the decline of its corn grinding counterpart. Steam pumps came into use quite early in the 19th century. Many drainage mills continued to work until the 1930s. They gradually decreased in number and although one was built as late as 1912 we can no longer number them in hundreds.

Some fine looking mills remain, however, among them Stracey Arms, near Acle Bridge; and High Mill, Berney Arms, Norfolk (preserved). The latter mill is a very interesting structure. Its scoop wheel is detached from the mill, and a long shaft provides a link between them. This was one of the few drainage mills which performed a milling operation—grinding cement clinker. Other drainage mills in various places were fitted with stones for corn grinding, but mills of this type were the exception rather than the rule.

Even when a pumping mill ceases work its shell can make a useful dwelling if it is not too remote. In the recent past the lonely Toft Monks mill, Norfolk, was offered for sale. A prospective buyer would have to be a keen waterman as this mill is not accessible overland, and the Broadland can be bleak and uninviting in the throes of an easterly January gale.

HYBRIDS

Like species in the natural world, windmills have also produced their hybrids. There were a few mills which were half post and half tower. The reason for their existence may be explained by a reluctance to demolish the sound body of a post mill when its working life could be extended by placing it upon a short tower. In such cases the gabled framework was mounted on a squat tower which had a curb like a conventional tower mill. This allowed the whole body to be luffed into the wind as conditions required. A fantail—above the rear gable—and a tiller beam of the usual post mill pattern provided alternative methods of moving the sails into the wind. Every rule has its exceptions, of course, and the composite mill at Monk Soham, Suffolk, was in fact a purpose-built structure and not the result of remodelling. Composite mills are also recorded at: Banham and Thornham, Norfolk; Rishingales, Suffolk and Cowick, Yorks.

Another variation on the mill design was to place a reduced post mill body on to a smock tower. The driving power was transferred from the windshaft by a vertical post, which was connected to the other machinery below. Not many of these hollow post mills seem to have been built in England, and the one on Wimbledon Common is a rebuilt survivor. The lower part of this unusual mill has been converted into a residence. Its sails were in proportion to the post mill style cap. This adjusted itself to changes in wind direction by the fantail—situated on an outrigged frame behind the rear gable. The original building is said to date from 1817, and its present appearance results from a rebuilding in 1890. Apart from its technical distinctions the mill house has an interesting literary association. Lord Robert Baden-Powell wrote part of *Scouting for Boys* within its walls in 1907. Two original hollow post mills remain at Stodmarsh, Kent and Acle, Norfolk.

CAPS

The Mill Cap

The crowning glory of the tower mill was its cap and sails. Over the years a good many disused mills have lost their caps. Some were purposely removed, others decayed and collapsed, or were forcibly removed by winter gales. Next to the sails the cap and its fittings is the most vulnerable feature of a mill. Cap design can tell us so much about a mill and it is a vital detail to record.

Some writers have suggested that certain styles of caps are common to given geographical areas. To some extent this is true but any general pattern of distribution tended to be affected by migrant millwrights, who sometimes went a long way from their homes to undertake work, and as a result produced a mill with features that were different from those of its immediate fellows. An example of this sort occurred at Much Hadham, Herts, where a millwright from Louth, Lincs, constructed an elegant eight sailer.

The function of the cap was to protect the tower and the machinery immediately below its roof. One of the most difficult areas to secure against the elements was the weather beam and forward end of the windshaft. Another important sector was the curb on which the cap revolved. This was usually guarded by a deep board or in the case of conical caps a neat and deep petticoat tastefully scalloped.

We can distinguish six main types of cap which may be found on smock and tower mills alike. They are: the gabled, the post mill type, the cone shaped, the domed, the ogee and the up-turned boat shape. It would be difficult to assign any definite periods of time to these different styles. They probably represent no more than a variety of solutions to the same technical problems.

The Gable Shape

One of the simplest forms of cap is the triangular gable and its variations. Examples of this kind could be found in widely scattered parts of England. They appeared as far north as the Fylde, Lancs; Wirral, Merseyside; Lutton Gouts, Lincs and Gwynedd. In the south they could be found at Bembridge, I.O.W.; Holmer Green, Bucks; and at Ashton, Somerset. It is interesting to note that what is probably England's only thatched mill—at High Ham, Somerset—also has a cap with contours that follow the swept gable pattern.

The Post Mill Shape

Many mills in Kent, Essex, Surrey and Sussex had caps shaped like miniature post mills. A sketch dated 1796 shows a Lincolnshire mill with a cap of this type, and the design may have been more general than present evidence suggests. Notable examples of this are at Headcorn, Canterbury and Cranbrook, Kent; Rye, Chailey, Staplecross and West Chiltington, Sussex. Caps of this kind do vary considerably in their proportions. Some are very deep and appear to sit heavily on their towers, others have a sleeker look and seem almost to float on air.

The Boat Shape

Another, East Anglian, type of cap has a remarkable similarity to an upturned dinghy. This style is related to the post mill caps but with one important difference. In plan its base timbers, instead of being parallel, are curved to match the contours of the upper part of the mill tower. Caps of this kind are in two forms: those with gables of almost equal proportions and those with a larger front gable and a more pronounced sweep to their ridge profiles.

Caps with balanced gables can still retain their individuality. If the cap is boarded in the usual way it can have quite a sleek look. An alternative can be found among those drainage mills where the weatherboarding is bound with additional strips running at right angles. This gives the impression of half a beer barrel, and has a much heavier appearance. Smock and tower mills may be found with boat shaped caps; and the most pleasing version visually is probably the one with the dramatic line to its ridge—like Gibraltar mill, Great Bardfield, Essex.

Conical Caps

The most elementary alternative to a conventional rectangular roof protecting a round tower, is to provide a cone shaped cap. This idea seems simple enough but it demanded a greater degree of skill to construct. Caps of this kind do not seem to have been as common as the other types—perhaps because they were re-designed when they required extensive repairs. Conical caps are either based on a circular plan or on a hexagonal or octagonal plan, as on the smock tower of Belloc's Mill at Shipley in Sussex. It had a burnt tiled roof. Another mill with an almost identical cap may be found at Cley-next-the-Sea, Norfolk.

Domed Caps

A dome is a logical development from a cone or pyramid, and its construction demands a considerable degree of skill. In Sussex they were shaped rather after the fashion of bee-hives and their proportions were light in appearance. Examples of typical Sussex domes are Polegate and Selsey. One at Patcham had the roundest of all.

In East Anglia the caps mostly sit heavily on squat towers —like deep bonnets on rotund matrons. The Cambridgeshire mill at West Wratting is typically heavy in appearance. Although the Suffolk smock mill at Alderton had a taller tower than most, it too boasted a large bonnet. It is the cap which gives the mills in this part of England their distinctive look and makes them readily distinguishable from their counterparts in the south and even those further west in Bedfordshire. There the Sharnbrook mill has a domed cap; but although it betrays certain East Anglian characteristics its cap has been diminished to the point where it complements rather than dominates the tower.

Going westward, Wilton mill, Wilts., sports a dome. It comes in the East Anglian class, but the profile is rounder and it gives an elephantine impression. This cap must have been very difficult to turn by hand.

The Ogee Cap

The ogee curved dome is the final glory of tower mill design. Among the ogee caps of England, those of Lincolnshire justly claim pride of place. Here they abound *par excellence* and in one form or another their influence extends across the eastern and midland parts of England. Sussex, too, has a few caps that are rather blunt specimens of the ogee style. The most noted one, Halnaker, was recorded by William Turner and his painting may still be seen in the National Gallery of Scotland.

Another example is the mill at Ewhurst, Surrey. Lurid tales are woven around this mill's working days when it appears to have been prominently involved in the distribution of contraband on its way from the coast to London. After the mill ceased work it was converted into a residence.

North of the Thames is the real ogee land, and many fine examples lie between the Thames Valley and Yorkshire. Out in the open Oxfordshire countryside—at Milton Common and Wheatley—we may see examples of the low pitched ogee cap which is picturesque without being elegant. Some caps in this part of the world have a copper cladding which weathers a distinctive green. Across the Oxfordshire border limestone mills are not unusual and in addition to the two mentioned above we may add Blackthorn—which has a tile hung tower.

At Biggleswade, Beds, the county's tallest mill has a definite Lincolnshire look. This is not surprising when we note that it was built in 1860 at a time when windmilling was at or near its apogee.

Two distinctions remain to be made about ogee caps. Those in the Cambridgeshire region—as at Burwell—have a slightly flared emphasis at the base, and in Lincolnshire the fashion is generally in favour of an inward curve which provided an unmistakable onion look. The details of finial decoration varies, but a round knob seems to be the favourite. South Midland finials tend to be heavier in design, and in Oxfordshire an acorn style is used.

Materials

Millwrights have employed many different materials in their efforts to weatherproof their mills. We may guess that the first caps on the early towers were thatched, but they could equally well have been boarded. Although corrugated iron adorns many a cap these days, millwrights have made good use of a variety of materials in the past. Chesterton, Warwicks, is leaded; copper was popular in Oxfordshire but perhaps the most universal alternative to these was tarred canvas on boards (called marouflage finish) which has many enduring qualities, and can be very attractive if it is skilfully applied— as at Heckington. Lincs.

SAILS

Common Sails

Early windmills had very simple rectangular wooden frames upon which the miller arranged his canvas. Sails were constructed around two sail stocks set at right angles and mortised through the wind shaft. Mills constructed in later times, *circa* 17th century, had the lengths of the sails increased by the addition of whips (lighter timbers bolted to the sail arms). An important difference between the primitive sails and those in use in the early part of the 18th century is shown in the diagram (p. 20). The canvas on the latter was rigged on a single sided blade in place of the earlier two sided arrangement. One of the important developments in windmill design came about in the 1750s when John Smeaton introduced the use of cast iron to supplement the traditional timber. When this happened it became possible to design cross arms and canisters to take stouter sail stocks; and iron gears were introduced to replace or complement existing wooden cogs. The use of iron gears running next to wooden wheels was said to make for quieter running and less vibration. The great disadvantage of common sails was the necessity to stop the mill working when the wind-strength changed so that the area of canvas could be adjusted. This was a tiresome and, in winter or stormy weather, a dangerous task.

Spring Sails

In 1772 a Scottish millwright—Andrew Meikle—invented a new type of sail which was composed of a series of shutters arranged like a Venetian blind. The shutter blades were opened or closed depending upon the state of the wind and how much work the miller had to do. In order to adjust the sails it was still necessary to stop the mill so that the bars connecting the rows of shutters on each sweep could be moved to the required position. Each sail was still altered independently, but the adjustment was a much easier matter than dealing with freezing canvas while clinging to a sweep high above the ground. The angle of each row of shutters was controlled by the pressure, transmitted by the shutter bar, from a powerful spring. Sails of this type were known as spring sails, and many mills had double rows of shutters on their sweeps.

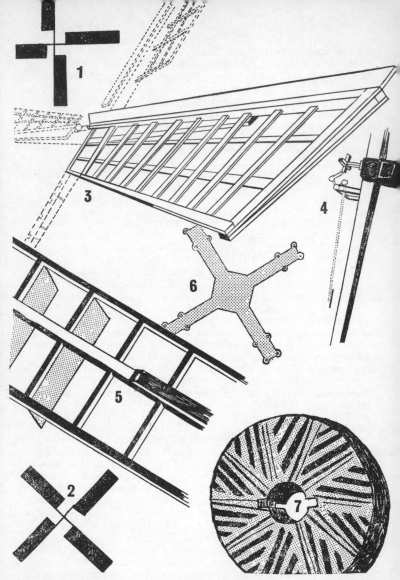

Sails: *1. Clockwise sail. 2. Anticlockwise sail. 3. Common Sail. 4. Sailstock with spider, bell crank and rod to control patent sails. 5. Rear view of patent sail frame with open shutter. 6. Iron cross—used in place of a canister. 7. Millstone erected as an ornament.*

Roller Reefing Sails

A variation of Andrew Meikle's system was invented in 1789 by Capt. Stephen Hooper. He replaced the shutters with a roller blind system which was claimed to be automatic, but in practice was not very effective. Mills which used this kind of sail often reverted to an alternative method when the rollers needed repair or replacement.

Patent Sails

The most significant improvement in sail design came about in 1807 when William Cubbit invented his patent sail. This retained Meikle's shutters, but they were controlled automatically by a weight suspended outside the mill. This weight was connected to the shutter bar by a series of rods linked, at the junction of the sails, to a lever known as the spider. Like spring sails, the patent shutters were frequently arranged in double rows; and their use was combined on some mills with a pair of common sweeps—as at West Wratting, Cambs. A later advance combined the spring device with a patent control to make spring sails fully automatic as well.

Multi-Sailed Mills

Not all mills had four sails—a good many managed in an emergency to work with two. The object of adding extra arms was to increase the effective area of the sails, and with it the efficiency of the machine. There were some mills built with five, six or even eight sails. The five-sailed mills were the least useful because the loss of one sweep rendered the mill idle until it was repaired. With one sail out of action the whole balance of the sails was lost. The last working five-sailer appears to have been the tower mill at Alford, Lincs; there was once a smock mill at Sandhurst, Kent, with a similar number of sweeps but it was demolished long ago.

Six sails provided a better selection of alternatives. If one sweep became unusable its opposite partner could be put out of use and the balance was restored. Even three sails out of order did not necessarily mean that the mill had to stop work, as alternate sweeps could be employed to maintain the balance. There were at least two Sussex post mills with six sails—at Hastings and Ashcombe (near Lewes). The latter was blown down in 1916. Further north, towers at Coleby Heath and Peterborough displayed the same sail arrangement.

According to Mr. Rex Wailes in *The English Windmill* there were but seven eight-sailers in England—at Diss, Norfolk; Eye, Cambs; Wisbech, Cambs; Much Hadham,

21

Herts; Holbeach, Rasen and Heckington, Lincs. This latter mill, the sole survivor, has been preserved by the Kesteven County Council.

The Annular Sail (The Wind Wheel)

The tall tower at Haverhill, West Suffolk, had a vast wheel instead of the conventional form of sail. Around the perimeter were ranged 120 shutters—each one five feet in length. Access to the shutters was provided by an unusually high gallery. This mill had a high domed cap and a fantail. The upper part of the wheel was over 80 feet above the ground, and it is not surprising to find that it could easily be seen from the neighbouring counties of Cambridgeshire and Essex. There was once another mill with this type of sail in that part of the world—it stood at Roxwell, Essex, and was dismantled in 1897. Windwheels of this kind may still be found at work in the Mediterranean—on Majorca.

A much smaller version of the Haverhill mill was once erected on top of a Sussex barn at Angmering. It operated a water pump, a turnip chopper and a corn mill. It ceased work over 50 years ago and details of its features are sparse. There can be no doubt about its wind wheel however which is recorded in R. Thurston Hopkins' booklet *Windmills* published many years ago in Sussex.

GRINDING THE GRAIN

The path taken by the grain through the millstones is shown in the diagram opposite. Millers made good use of gravity long before it was 'discovered' by Isaac Newton. Corn was first taken, via the sack hoist, to the top of the mill where it was placed in the grain bin. A chute led to the hopper positioned above the millstones. Grain trickled from the bottom of the hopper on to the feed shoe which was methodically shaken by the rotating damsel or by the shaker on the crutch pole. As the upper runner stone revolved so a few grains at a time were fed into the eye to be ground and expelled around the stones' circumference. Then the meal fell into the meal spout and finally into the bin on the floor below. The effect of the grinding process was to make the emerging meal quite warm, and anyone feeling it for the first time usually expresses surprise at its temperature.

In their heyday most mills operated two types of stone. Barley was worked on Derbyshire Peak stones which were hard and suited to this type of grain. Flour was usually pro-

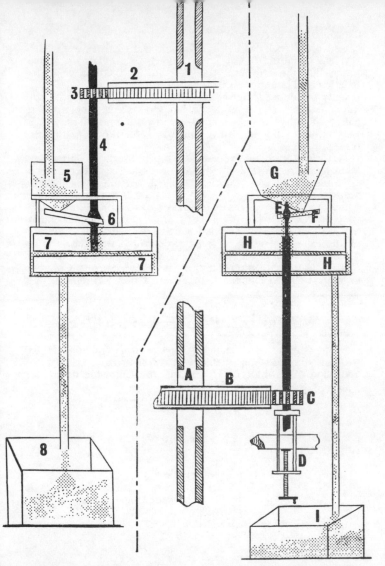

Corn Grinding Machinery: OVERDRIVEN: *1. Main Shaft. 2. Great Spur Wheel. 3. Stone Nut. 4. Crutch-Pole. 5. Hopper. 6. Feed-Shoe. 7. Mill Stone. 8. Meal Bin.* UNDERDRIVEN. *A. Main Shaft. B. Great Spur Wheel. C. Stone Nut. D. Jack Ring—to lift stone nut out of gear. E. Damsel. F. Feed Shoe. G. Hopper. H. Mill Stone. I. Meal Bin.*

cessed on French Burr stones which were harder and better for finer grinding. These latter stones were not made in one piece, like the Derbyshire Peaks, but in sections cemented together and bound with iron bands.

Only the upper (runner) stone revolved and it did not come into contact with the bedstone below as some people are inclined to think. The space between the surfaces of the stones was minute and carefully controlled to produce the best results.

Not all stones revolved in the same direction, and it was their mode of revolution that determined the pattern of the dressing on the grinding faces. Clockwise or anti-clockwise motion depended upon the way the sails revolved. Millstones which revolved clockwise were easier to dress and in the conventional post mill—with two pairs of stones in the head of the mill—this usually implied a counter clockwise revolution of the sails. An observer can decide which way the sails revolved by noting the position of the sailstock or whip in relation to the sail frame—the sailstock always leads. Common sails are easier specimens to ascertain if they are more or less intact.

WINDMILL FACTS & FIGURES

From Primrose McConnell's *Agricultural Notebook* (1883) we may glimpse some of the technicalities associated with the operation of a windmill. He records the following data:

Angle of shaft of sails with horizon
$\quad\quad\quad$ =8° on level ground up to 15° on exposed heights.

Length of sails $\quad$ =4 times the breadth.

Length of sails $\quad$ =6/7ths of length of arm or 'whip'.

Arm divides sails=proportion of 3 x 5; narrow part next the wind.

Area of sails $\quad$ =$\frac{1}{4}$ area of circle.

Area of sails $\quad$ =$\frac{1}{8}$ of area of part occupied by vanes in small self-regulating windmills; this gives greatest effect.

Angle of sails to plane of motion
$\quad\quad\quad$ =5° at tip up to 22° next the axis.

Revolutions of mill stone
$\quad\quad\quad$ =5 to 1 of sails.

Revolutions of sails
$\quad\quad\quad$ =12 per minute with the wind at a velocity of 20 ft. per second.

1. The post mill at Mountnessing in Essex.

2. The Cat and Fiddle post mill near Dale Abbey, Derbyshire.

3. The tower mill at Great Bardfield in Essex.

4. Detail of the sails of the tower mill at Burgh-le-Marsh, Lincolnshire.

5. The spur wheel and quant post at Burgh-le-Marsh.

6. *The fantail of the post mill at Saxtead Green, Suffolk.*

7. *The tower mill at Morcott in Leicestershire has an ogee cap.*

8. The tower mill at Cley-next-the-Sea, Norfolk.

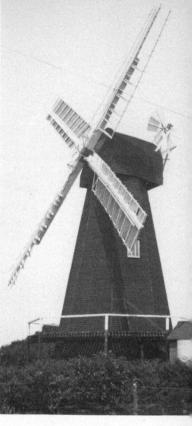

9. Stelling Minnis smock mill, Kent.

10. The preserved drainage mill at Thurne Dyke in Norfolk.

11. White smock mill near Sandwich, Kent.

12. Billingford tower mill, Norfolk.

13. *England's last eight-sail mill, at Heckington, Lincolnshire.*

14. *The unusual tower mill at Chesterton, Warwickshire.*

15. Left: The thatched windmill at High Ham in Somerset.

16. Above: Pitstone post mill in Buckinghamshire is the oldest dated mill in England.

17. The drainage mill at Wicken Fen in Cambridgeshire is a miniature smock mill.

18. *Wilton tower mill, near Great Bedwyn in Wiltshire, has now been restored to working order.*

19. *Below: restoration work being carried out on Friston post mill, Suffolk. The work is now complete and the mill open to visitors.*

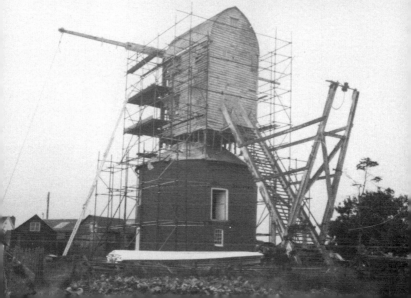

20. *One of the few working windmills left, this is the tower mill at Pakenham, Suffolk.*

SOME NOTABLE WINDMILLS

In this section a few of the more interesting mills in the counties listed are described. The list is not exhaustive and complements the Gazetteer. Ownership is indicated by the name given in brackets at the end of each entry.

BEDFORDSHIRE

Stevington—post mill (1770): This is the finest mill in the county to survive the last war. It has two features worth noting—a mansard roof, and an unusually prominent flared skirt. This may have been the last working mill with common sails. It has, understandably, attracted the attention of the artist. The mill stands on the Bedford side of the village. Park at the bottom of Mill Lane and do not tread in the growing corn which surrounds the mill during the summer months. An illustrated booklet can be purchased from the village stores. (Bedfordshire County Council.)

BUCKINGHAMSHIRE

Brill—post mill (1668): This fine mill commands a view across the Vale of Oxford to the Cotswolds. Earlier in this century it had an open trestle which is now protected by a brick roundhouse. On the exposed hillside at Brill there were once three mills. An etching (1926) by Ian Strang may be seen in the County Museum, Aylesbury.

Lacey Green—smock mill (c. 1650): This mill is certainly the oldest remaining example of its type, and its massive wooden machinery represents a significant part of our technological history. A complete restoration is being carried out by the Chiltern Society. Open to the public on Sunday afternoons from May until October.

New Bradwell—tower mill: This mill has been restored by the Milton Keynes Development Corporation. The sails were replaced in 1975. The mill has a new brake wheel and wallower made by Mr J. Davies of Lincolnshire. Viewing by appointment: telephone Milton Keynes 71234 (Conservation Department) for details.

Pitstone Green—post mill (1627): This is the oldest dated mill in England—see page 5 above. It stands on the south side of Ivinghoe by the B488 road which leads to Tring. On the west side of the road (the righthand side as you go towards Tring) where it bends sharply there is room to park. The mill is clearly visible from this point—in the field beyond the stile. Open to the public on Sunday afternoons and Bank

Holidays from May until October. Parties at other times by arrangement. (National Trust).

Quainton—tower mill (1830): This is the tallest tower mill in the county. It is in a fine position at the top of the village green overlooking the Vale of Aylesbury. The mill is being restored and it is open to the public as advertised.

Wendover—tower mill (1796): Wendover mill's claim to distinction lies in the fact that it has the largest cap in the land. The hexagonal tower (66ft high) is 25 feet across at the base—where the walls are 3 feet thick—and 18 feet wide at the curb. An ingenious winch worked by a worm gear once allowed the cap to be turned from the inside. In 1904 the sails were removed—following storm damage—and a steam engine was employed to power the stones. During the General Strike (1926) the mill ceased work. It became a residence in 1931. No machinery survives. (Private).

CAMBRIDGESHIRE

Bourn—post mill (15th cent?): Although Pitstone Green mill has the distinction of being the oldest dated mill it cannot be said to be typical of the earliest post mills. Bourn mill alone reflects the pattern of the earliest mills known to us from carvings and manuscript illustrations. This mill is known to have occupied its present site since 1636, and it is probably the country's real veteran. (Cambridge Preservation Society).

Great Chishill—post mill (18th cent.?): A splendid open-trestle mill on the west side of the village, it stands on an open site near a row of cottages where the road starts to dip downhill. The carefully restored boarding is painted white which makes an interesting contrast to such mills as Pitstone or Bourn. The spider testifies to the former presence of shutters. Above the mill steps the fantail stands proudly upon its carriage—another unusual feature in this part of the countryside. In 1882, when the parish was a part of Essex, the miller was Henry Andrews. An idyllic location.

Histon—smock mill: An attractive mill in a typical village setting. Restored by the previous owner. Has a fan and sail stocks. (Private).

Madingley—composite mill: It was formerly a post mill, situated at Ellington, and rebuilt in 1936 by C. J. Ison (of Histon). It stands on the site of a previous mill which was blown down in 1909. Situated at the side of the busy A45—parking severely restricted. (Private).

Wicken Fen—drainage mill: Part of the extensive nature sanctuary which attracts so many visitors. 10 miles N.E. of

Cambridge, in a beautiful setting. The white sails stand out as you approach from Stretham. Look for the National Trust sign as you enter the village and then turn (right if you come from Stretham) down the narrow lane to the Fen. This delightful little mill was moved from Adventurers Fen (part of Burwell Fen) and was originally known as Normans Mill. A plate records the details. Notice the position of the scoop wheel and the tailpole to rotate the minuscule cap. (National Trust).

ESSEX

Bocking—post mill: This stands close to Bocking Church Street. The mill was moved from a site about a quarter of a mile away in 1830. The roundhouse contains an interesting array of milling relics. Messrs. Noble & Sons of Ongar completed the restoration work in 1964. Keys available from The Bull inn.

Stansted Mountfitchet—tower mill: The mill, now scheduled an ancient monument, was built by Joseph Linsell in 1787, and was then continuously in service until 1910. Subsequently the mill was conveyed for the benefit of the 'inhabitants of Stansted', Mr. Rex Wailes and the Parish Council being appointed co-trustees. In 1966 an extended restoration programme was undertaken, replacement of the sails structure and the fantail being included. The internal machinery and power transmission are retained in complete form. The Mill is normally open-to-view between 2.30 and 7.00 p.m. on the first Sunday of each month from April to October and on Sundays and Mondays at the Easter, Spring, and Summer Bank Holiday week-ends. For arrangements to view at other times telephone Stansted 3596. Admission charge. A pictorial guide is also available. Further details from: P. R. Gibson, Stansted Mountfitchet Windmill Committee, 46 Wetherfield, Bentfield End, Stansted, Essex.

HEREFORD AND WORCESTER

Avoncroft Museum of Buildings, Stoke Prior, Bromsgrove (telephone: Bromsgrove 31363): One of the Museum's most important exhibits is the re-erected post mill from Danzey Green, near Tanworth-in-Arden.

HERTFORDSHIRE

Cromer—post mill (17th cent.): Since 1966 this mill has been gradually restored by the Herts. Building Preservation Trust Ltd., and donations are still required by The Treasurer H.B.P.T. The Chantry, 69 Sish Lane, Stevenage, Herts. The mill body, which has an ogee shaped gable, has been reboarded and painted, and the sails, steps and fan have been restored. The trestle is protected by an octagonal round house. In 1926 the building was damaged in a gale and it then lost a sail which caused it to become headsick. This prevented the sweeps from rotating. At this time there were two pairs of stones—in head and tail. The iron windshaft was formerly driven by double patent sails. The mainpost has been called "the largest piece of oak in Hertfordshire"—D. Smith op. cit. p 110. Cromer mill stands close to the roadside on the Buntingford side of the village.

HUMBERSIDE

Wrawby—post mill (c. 1790): This mill was carefully restored (1961-5) by a group of local enthusiasts with the assistance of Thompsons of Alford. There are two pairs of stones —in head and tail. The French stones (head) are 4ft. 8ins. dia. and the Peak stones are 4ft. 2ins. dia. Coil springs control the sail blades (called shades in this part of the country) and the sails are mounted on an iron cross arm. There is a collection of milling bygones in the roundhouse. A descriptive leaflet is available. Wrawby mill ceased work commercially in 1940 when it lost two sweeps. From May to August (including Bank Holidays) the mill is open on Saturdays and Sundays— 1.30 to 7.0 p.m. Admission charge. Bags of stone-ground wholemeal flour can be purchased. Private visits by individuals or groups can be arranged at other times on written application to the Custodian—Jon A. Sass, St. David's, Wrawby Road, Brigg, South Humberside.

ISLE OF WIGHT

Bembridge—tower mill (c. 1700): Given to the National Trust by Mrs. E. Smith and now restored. The mill was last worked in 1913 and a good deal of its original machinery remains. Open Spring Bank Holiday—Sept. 30, every day 11—6. At other times by appointment with National Trust Office, 35a St James Street, Newport, Isle of Wight.

KENT

Chillenden—post mill (1868): An open trestle mill with four sweeps. It replaces an older post mill dating from the seventeenth century. This attractive mill is about half a mile north of the church. (Kent County Council).

Cranbrook—smock mill (1814): This must be the most glorious mill in the country. Rising up for some 75 feet above the Kentish landscape it presides over the town with a unique air of paternalism. Shuttered sails arc across the sky a hundred feet or so above the spectator's head, and add elegance to its unequalled dignity. Here is a place of pilgrimage for the windmillian. The visual splendour afforded by this masterpiece owes much to the fine proportions defined so long ago by its builder James Humphrey: truly a mill not to be missed. (Private).

Meopham—smock mill (1801): A hexagonal mill with five floors, standing on the village green. Extensively repaired in 1962, by E. Hole and Son of Burgess Hill, Sussex, for the County Council at a cost of some £4,300. This mill is now subjected to an annual inspection. Its sails are rotated a quarter of a turn every three months. (Kent County Council).

Sandwich (White Mill)—smock mill (c. 1760): Restoration work on this mill began in 1961 and has proceeded as funds permitted. One cant post and parts of two others were replaced. Other work includes new weatherboarding and two sweeps (fitted 1967). The mill will idle in a moderate wind. Its machinery is intact. It last worked by wind in 1926 and then with an oil engine until 1953. Grants for repairs were made by Kent County Council, S.P.A.B. and local subscriptions. Vincent G. Pargeter, millwright, carried out the repairs listed. (Private).

LEICESTERSHIRE

Wymondham—tower mill (19th cent?): An interesting mill with an elegant cap. Not well-recorded in the past but it has a claim to special interest as it once displayed six sails—the cross arm remains. At the side of the mill the rectangular steam-engine house survives and the pulley wheel projecting from the mill tower shows us how the steam power was transmitted to the machinery. (A private mill—accessible with permission from the owner).

LINCOLNSHIRE

Alford—tower mill: A five-sailed mill in working order which is still used. Visitors are admitted and a descriptive leaflet is available. A mill not to be missed even in a county renowned for its magnificent windmills.

Boston—tower mill (19th cent.): The Maud Foster mill takes its name from the Maud Foster Drain which it overlooks. Next to the "Stump" this is one of the most attractive and prominent features of the Boston skyline. Its tower is constructed from attractive yellow bricks and the immediate surroundings have changed very little since Stanley Freese sketched this five-sailer in 1931—see *In Search of English Windmills*. Since then the blades of the fantail have gone but the

45

frame remains. Although the sails have lost their shutters the outline of the sweeps can still be appreciated. (Private).

Burgh-le-Marsh—tower mills (19th cent.): There were once two mills here and the derelict tower, above the church, should not be confused with the outstanding five-sailer at the bottom of the hill. This latter mill (built 1833) was purchased by the county council in 1965 and has been carefully restored to working order. Its slender tarred brick tower is crowned with a typical Lincolnshire white cap. The five patent sails are interesting numerically and by their clockwise rotation. There are four floors, the upper one is normally closed but open on request. Visitors should spend most time on the second floor where the casing of one pair of stones has been cut away to give a side view of the assembly—see plate 5. The runner stone of another pair has been removed to show how the quant post (crutch pole) fits into the mace. The stones on the first floor were driven by a diesel engine housed outside the tower. A meal mixer, a crusher, scales and a sack barrow can be seen on the ground floor. The last miller was Mr. Dobson and the mill is often called Dobson's Mill. Admission is free *at your own risk* but unaccompanied children are not given access to the mill. The key is available from the adjacent Windmill Restaurant where postcards and an informative leaflet can be obtained. Good car-parking facilities. (Lincolnshire County Council).

Heckington—tower mill (1830): This is England's last eight-sailer. It was originally constructed in the early nineteenth century but its present cap and sails were brought here from Boston in 1892—by John Pockington. Rising high above the landscape its majestic tarred tower dominates the scene and no photograph will do justice to the subtly curved shadows the summer sun casts upon the tower walls. Long ago it also worked a saw-mill as well as its five sets of stones. The mill was purchased by the county council in 1953 and then restored to its present glory. Visitors are welcome and the key is available—at reasonable times—from (another) Mr. John Pockington at the adjacent mill house. (Lincolnshire County Council).

Sibsey—tower mill (1877): In the last edition of this book Sibsey mill was called a 'lovely tatterdemalion'. It now has a new cap and all six sails. Other internal repairs are also being made. Trader Mill, as it is locally named, possesses the finest wrought iron gallery in England. Rex Wailes describes the interior in some detail. Leave Boston by the A16 and then take the B1184 (Frithville) road, off to the left, as you enter Sibsey.

LONDON

Brixton Mill, Blenheim Gardens, Lambeth—tower mill (1816): The tall tower, with its spirally arranged windows, has a boat shaped cap. This mill was renovated by the L.C.C. in 1964. Some of its present machinery comes from the defunct tower mill at Burgh-le-Marsh, Lincs., and the remainder (plus the new sails) were newly made by Thompson & Son, Millwrights, Alford, Lincs. G.L.C. leaflet (Publication 123) available from County Hall, S.E.1. (Greater London Council).

Upminster, St. Mary's Lane—smock mill (c. 1800): One of the best mills close to London. When a new sail stock was inserted earlier this century the cost amounted to £108. Not long afterwards the fantail collapsed and involved the miller in another bill of £120. When the Havering Borough Council renovated this fine mill not so long ago costs had risen rather dramatically but enthusiasts will readily acknowledge the Council's wisdom. A good deal of the machinery remains. Mr. Alfred Abraham was the miller (in 1931) and he followed his father and grandfather. A scale model of this mill, made by the late E. W. King, may be seen in the museum section of Romford Central Library.

Wandsworth Common—pumping mill (1837): The tower survives at the junction of Windmill Road and Spencer Park Road. See: K. G. Farries and M. T. Mason—op. cit. p. 214.

NORFOLK

Berney Arms, near Gt. Yarmouth—drainage mill (c. 1860): Its 70 feet tower is said to be the highest in East Anglia. It was restored in 1967 as a National Monument by the Ministry of Public Building and Works. This mill served a double purpose as it could also grind clinker—the scoop wheel is detached from the tower. Rex Wailes—op. cit. p. 76—gives a good account of its features and provides an excellent drawing of its technical details. The only access is by rail or water. Open during summer months only—9.30 to 7.0 p.m. Leaflet and postcards available. (Department of the Environment).

Billingford—tower mill: The key to the mill is available from the Three Horseshoes. No charge but contributions welcome.

Burnham Overy—tower mill (1816): Constructed to complement the adjacent water mill. The mill became tail-winded in a storm during 1914. It was eventually reconditioned and converted to a dwelling in (1926) by Mr. and Mrs. H. C. Hughes who presented it to the National Trust in 1958. Not open.

Horsey Drainage Mill—tower mill (1912): The original tarred tower mill (once known as Horsey Black Mill) probably dated from the early eighteenth century. In 1895 it became tail-winded and lost its cap. By 1912 the structure became very shaky and was dismantled to its foundations. At the time of rebuilding the vertical shaft was replaced by one in Scandinavian pine but the other machinery is largely original. Hornbeam was used for the cogging. The maximum lift was 7 feet. This mill continued work until it was severely damaged by lightning (1943) which split the sail stocks. The difficulty in obtaining timber at that time prevented their replacement and the stocks were removed in 1956. The mill's present state of preservation is due to the joint efforts of the S.P.A.B. and the Norfolk C.C. Since 1948 the mill has been owned by National Trust. During summer months it is left open daily for visitors who may explore the interior—*at their own risk*. Leaflet available from N.T. Area Office, Blickling, Norwich, Norfolk.

OXFORDSHIRE

Milton Common—tower mill (18th cent.): A mill to be viewed against the slanting rays of a sunrise or sunset if the spectator is to gain most from the prospect. Although it has no claim to technical distinction its setting makes it a highly significant landscape feature—perched on an eminence over-looking a limb of the Thames Valley. The round tower with its sash windows declared its period of origin. The remains of the sweeps show that they were canvas rigged. The mill was given a new cap in 1976. (Private).

SOMERSET

Ashton Mill, Chapel Allerton—tower mill (19th cent?): There were once about 50 mills in the county and this is the sole survivor in a complete state. Like its fellows it lost its flour trade to the larger mills at Avonmouth. It ceased work—as a grist mill—in 1927. In the late 1950s it was purchased by Mr. C. C. Clarke who provided for its entire restoration. He presented the mill and its field to the Bristol City Museum in 1966. Ashton Mill is the finest mill to be found in western England.

High Ham—tower mill: England's only stone and thatched windmill. It was built in the 1820s and ceased work in 1910. This mill is located on a hilltop not far from Langport. It is in a delightful setting next to the mill house which has a beautifully kept garden. The property is owned by the National Trust and visitors are admitted at the times stated in the Trust's literature. Appointments to view can be made by writing to the custodian—Mrs. K. D. Taylor, The Mill, High Ham, Langport, Somerset. An illustrated leaflet is available. (National Trust).

SUFFOLK

Friston—post mill (c. 1812): This is probably the tallest post mill in England. It is 55 feet to the ridge. In its working days it ground a ton an hour and employed three millers. It has a roundhouse of two storeys and a fan carriage of inspiring proportions. The slender buck houses three pairs of stones. In the rear gable there is an unusual Gothic window. The restoration of this mill has taken several years but the result is quite magnificent. Open to the public as advertised.

Herringfleet—smock mill: Built by Robert Barnes in 1830 and restored by the Ministry of Works and East Suffolk County Council. This is a drainage mill and the couch and stove once used by the marshman who kept the mill working can still be seen. Nightwork was essential during periods of heavy rain. The sails are still spread with canvas when large parties of visitors are entertained. A booklet is available from the County Offices at Ipswich.

Holton—post mill (c. 1749): Open to the public on Spring and Late Summer Bank Holidays.

Pakenham—tower mill: Another of the few working windmills. The tower is about 80 feet tall. Forty gallons of tar are required to cover the walls and fourteen gallons of white paint are used on the sails and cap. An illustrated booklet and postcards are available at the mill.

Saxtead Green—post mill: This is a splendid restoraion of a privately owned corn mill (by Ministry of Works in 1957-60). It was producing flour up to the 1914-18 War and cattle feed until the owner died in 1947. The mill is in operating condition including the fantail which automatically turns the mill to face the wind. Open daily except Sundays. Leaflet. books and postcards available adjacent to mill or from present owner at mill house.

Sutton—tower mill (1789): A typical red-brick tower. The sails were 73 feet across and contained 216 shutters. An RAC sign directs visitors to the mill which is open from March to October (9.30—6). Admission charge. A guide to the mill is also available.

SURREY

Outwood—post mill (1665): This is the oldest working mill in the country. It has spring sails controlled by elliptical springs. Extensive repairs were carried out in 1952 which accounts for the mill's present fine appearance. (Private).

SUSSEX, EAST

Cross-in-Hand—post mill: This mill had been moved twice (in 1855 and 1856) before it reached its present site. It originally stood at Mount Ephraim, Uckfield; see Peter Hemming—op. cit. p. 54. An unusual feature is the largely metalled body. There is a fine fan carriage on the tailpole. The roundhouse has two storeys—of tarred brick and galvanized iron. On 5th June 1969 the mill lost a sweep. As it fell it split another sailstock.

Nutley—post mill (c. 1670): This interesting open trestle mill has a rather tall slender body. It has not worked for about 60 years when it operated with spring set sails. Sweeps were fitted in the 1930s to preserve its appearance. The Nutley Windmill Appeal (Uckfield and Dist. Preservation Trust) has been set up in order to raise the £3500 required to ensure the mill's future. Its machinery is intact and the two pairs of stones are unusually arranged with the burrs in the head and the peaks in the tail. The mill may be reached from the Nutley —Crowborough road about a mile from Nutley. (Private).

Polegate—tower mill (1817): All internal machinery remains (restored 1967) and the storeroom contains a collection of milling bygones. Open from May to October inclusive on Sundays, Spring and Summer Bank Holidays, and all Wednesdays in August, 2.30—5.30 p.m. Parties by arrangement, apply D. Jones, 22 Manor Road, Hampden Park, Eastbourne. Tel: Eastbourne 54845. The sails turn when the weather allows. (Property of Eastbourne and District Preservation Trust).

SUSSEX, WEST

Belloc's Mill, Shipley—smock mill (1879): This is the largest mill in the county, and among its technical features we should note the double shuttered sails, three pairs of stones and an octagonal cap. The mill was purchased in 1906 by the writer Hillaire Belloc who lived there and kept the mill working. After Belloc's death (1953) the mill decayed and was in a poor condition when it was restored in 1957 as a memorial to him. It is open on the first Saturday and Sunday of each month and worked if there is sufficient wind. Open 2.30 p.m. to 5.45 p.m. It is possible to arrange for parties to visit the mill at other times by *written* appointment with Mr. R. Jebb, King's Land, Shipley, Horsham but casual visitors cannot be admitted. Admission charge. (Private).

Weald and Downland Open Air Museum, Singleton. Among the exhibits at this splendid museum of buildings is a rather primitive wind pump with four common sails. The museum is open as advertised.

WARWICKSHIRE

Chesterton—tower mill (1632): Formerly attributed to Inigo Jones. The tradition that this tower was built as an observatory and then converted to a mill is (probably) not correct. This elegant mill has recently been restored. Rex Wailes includes a detailed drawing—op. cit. p. 67. (Warwickshire County Council).

WILTSHIRE

Wilton—tower mill (1821): The last windmill in a more or less complete state to survive in the county, it stands on a hilltop about half a mile east of Wilton village, near Great Bedwyn (not to be confused with Wilton near Salisbury). The five-storey tower worked for about a century and it has now been restored to its former glory. There are two common and two patent sails. Open on Sundays from Easter to the end of September 2—5 p.m. Admission charge.

GAZETTEER

In the following county lists some of the mills which are still visible are recorded. They do not presume to provide an exhaustive account of remaining mills, and readers may care to suggest any distinctive mills which have been overlooked. Mills in rather isolated positions frequently have the name of a nearby town or village added to them; and the letter, in brackets, at the end of each entry indicates the type of mill. Thus: post mill= (p); smock mill (s); tower mill (t); composite mill (c) and some visible remains (x).

AVON
Brockley (t); Clifton Down, Bristol (t); Falfield (t); Felton Common (t); Frampton Cotterell (t); Hutton (t); Kenn (t); Locking (t); Portishead (t)—now part of golf club house; Uphill (t); Warmley (t); Worle Hill (t); Worle Vale (t).
BEDFORDSHIRE
Dunstable (t); Henlow (s); Holcot (t); Houghton Conquest (t); Houghton Regis (t); Lower Dean (p); Potton (t); Sharnbrook (t); Shefford (t); Stanbridge (t); Stevington (p); Thurleigh (t); Tottern-hoe (t)—Doolittle Mill, a combined windmill and watermill; Upper Dean (t); Woburn (s).

BUCKINGHAMSHIRE
Brill (p); Bradwell, New (t); Cholesbury (t); Coleshill (t); Fulmer (t); Great Horwood (x); Ibstone (s); Lacey Green (s); North Marston (x); Pitstone Green (p); Quainton (t); Stewkley (x); Thornborough (x); Wendover (t).

CAMBRIDGESHIRE
Arrington (t); Ashley (s); Barnack (t); Barrington (t); Bourn (p); Burwell 2 mills (t); Castor (t); Chatteris (s); Chippenham (s); Coates (t); Cottenham (t); Doddington (t); Eaton Socon (t); Elsworth (t); Ely (s); Fulbourn (s); Great Chishill (p); Great Gidding (t); Great Gransden (p); Guilden Morden (t); Haddenham (t); Harston (t); Hemingford Grey (t); Hildersham (t); Histon (s); Ickleton (t); Kneesworth (t); Linton (t); Little Street (t); Little Wilbraham (t); Madingley (p); March (s); Over (t); Pymore (t); Ramsey (s); Sawtry (s); Six Mile Bottom (p); Soham 3 mills (t, s, s); South Stanground (Farcet) (t); Steeple Morden (s); Streetly End (t); Stretham (t); Swaffham Prior (s); Swavesey (t); Thorney (t); Upwell (t); Upwood (t); Werrington (t); West Wratting (s); Weston Colville (s); Whittlesey (t); Wicken (s); Wicken Fen (s); Willingham (s); Wisbech (t); Woodditton (s).

CLEVELAND
Elwick (t); Hart (t).

CORNWALL
Carlyon (t); Mount Hermon (t).

CUMBRIA
Cardewlees (t); Cockermouth (t); Haverigg (t); Langrigg (t); Monkhill (t); Wigton (t); Workington (t).

DERBYSHIRE
Dale Abbey (Cat & Fiddle Mill) (p); Heage (t); Kirk Hallam (p); Normanton (x).

DORSET
Shaftesbury (t)—built 1969 in the Portuguese style.

DURHAM
Aycliffe (t); Easington (t); Ferryhill (t); Hutton Henry (t).

ESSEX
Ashdon (p); Aythorpe Roding (p); Bocking (p); Clavering 2 mills (t); Debden (t); Dunmow (t); Finchingfield (p); Gainsford End (Toppesfield) (t); Great Bardfield (t); Mountnessing (p); Ramsey (p); Stansted Mountfitchet (t); Terling (s); Thaxted (t); Tiptree (t); White Roding (t).

GUERNSEY, C.I.
St. Martin's (t).

GWYNEDD
Felin Adda (t); Felin Llanerchymedd (t); Llyn On, Llandeusant (t); Melin-y-Bont, Bryn Du (t)—a windmill with a waterwheel and a list of miller's charges in Welsh.

HAMPSHIRE
Bursledon (t); Chalton (t); Langstone (t); Portchester (t).

HEREFORD AND WORCESTER
Avoncroft Museum of Buildings, Bromsgrove: the post mill from Danzey Green.

HERTFORDSHIRE
Brent Pelham (s); Charlton (t); Colney Heath (t); Cromer (p); Croxley Green (t); Great Hormead (x); Great Offley (t); King's Walden (t); Little Hadham (t); Reed (t); Tring (t); Weston (t).

HUMBERSIDE
Bainton (t); Barton-on-Humber 4 mills (t); Beverley 2 mills (t); Bridlington (t); Burstwick (t); Ellerton (t); Goole (t); Hibaldstow (t); Howden (t); Keyingham (t); Kirton Lindsey (t); Nafferton (t); Ousefleet (t); Patrington (t); Scawby (t); Scunthorpe (t); Seaton Ross 2 mills (t); Skidby (t); Stallingborough (t); Swinefleet 2 mills (t); Waltham (t)—once 6 sails; Wrawby (p); Wressle (t); Yapham (t); Yokefleet (t).

ISLE OF WIGHT
Bembridge (c. 1700) (t).

KENT
Ashford (s); Benenden (s); Bidborough (t); Canterbury (1817) —St. Martin's Hill (t); Charing (s); Chillenden (1868) (p); Cranbrook (1814) (s); Chislet (1765) (s); Eastry (s); Edenbridge (t); Faversham (t); Guston (1849)—Swingate Mill (t); Herne (1781) (s); Kingston—Reed Mill (t); West Kingsdown (s); Kippings Cross (x); Margate—Draper's Mill (s); Meopham Green (1801) (s); Northbourne (1848)—New Mill (s); Oare (1878) (t); Preston (Sandwich) (x); Reculver (s); Ringwould—Ripple Mill (s); Rolvenden (p); Sandwich (s); Sarre (1820) (s); Stanford (1875) (t); Stelling Minnis (1866) (s); St. Margaret's Bay (1929) (s); Sutton Valence (x); Whitstable (1815)—Boarstall Hill (s); Willesborough (1869) (s); Wittersham (1781)—Stocks Mill (p); Woodchurch—Lower Mill (s).

The dates given for the Kent mills have been provided by Vincent G. Pargeter.

LANCASHIRE
Clifton (t); Kirkham (t); Lytham St. Annes (t); Staining (t); Thornton Cleveleys (t); Treales (t).

LEICESTERSHIRE
Kibworth Harcourt (p); Morcott (t); Waltham-on-the-Wolds (t); Whissendine (t); Wymondham (t)—once 6 sails.

LINCOLNSHIRE
Alford (t)—5 sails; Baston (t); Boston (Maud Foster mill) (t)— 5 sails; Bourne (x); Burgh-le-Marsh (t)—5 sails; Heapham (t); Heckington (t)—8 sails; Holbeach (t); Huttoft (t); Kirkby Green (t); Kirton End (t); Langtoft (t); Lincoln (t); Long Sutton (t)— 6 sails; Mablethorpe (x); Sibsey (t)—6 sails, 4 remain; Sleaford (t); Stickford (t); Stickney (t); Swineshead (t); Wainfleet (t); Wigtoft (t); Winthorpe (t); Wragby (t).

LONDON
Arkley (t); Brixton (t); Islington (t); Keston (p); Mitcham Common (x); Plumstead Common (t); Shirley (t); Upminster (s);

Wandsworth Common (s); Wimbledon Common (hollow post mill—originally).

MERSEYSIDE
Bidston (t).

NORFOLK
Berney Arms (Td)

Acle (t); Aslacton (t); Aylsham (t); Billingford (t); Blakeney (t); Burnham Overy (t); Carbrooke (t); Caston (t); Cley (t); Denver (t); Diss (t); East Dereham (t); East Harling (t); East Runton (t); East Ruston (t); East Wretham (t); Frettenham (t); Gooderstone (t); Garboldisham (p); Gayton (t); Hempnall (t); Hickling Green (t); Hindringham (t); Hingham (t); Holt (t); Honingham (t); Horning (St. Benet's Abbey) (t); Horsford (t); Harpley (t); Ingham (t); Little Cressingham (t); Little Melton (t); Ludham (High Mill) (t); Mileham (t); Mulbarton (t); Neatishead (t); North Creake (t); Norwich (Lakenham) (t); Old Buckenham (t); Overy Staithe (t); Paston (t); Potter Heigham (t); Ringstead (t); Rockland (t); Roughton (t); Scole (t); Sea Palling (t); Sedgeford (t); Shouldham Thorpe (t); Stokesby (t); Stoke Ferry (t); Stratton St. Michael (t); Sutton (t); Thurne (t); Walpole St. Peter (t); West Walton Highway (t); West Winch (t); Weybourne (t); Wicklewood (t); Worstead (t); Yaxham (t).

NORTHAMPTONSHIRE
Badby (x); Greens Norton (t); Hellidon (t).

NORTHUMBERLAND
Acomb (t); Bamburgh (t); Chollerton (t); Great Whittington (t); Haggerston (t); Hartley (t); High Callerton (t); Scremerston (t); ? Spindlestone (t); Woodhorn (t).

NOTTINGHAMSHIRE
North Leverton (t)—a working mill.

OXFORDSHIRE
Blackthorn (t, x); Bloxham (p); Milton Common (t); North Leigh (t); Wheatley (t).

SOMERSET
Chapel Allerton (t); Curry Rivel (t); High Ham (t); Shapwick (t); Stone Allerton (t); Walton (t); Watchfield (t); West Monkton (t).

SUFFOLK
Aldeburgh (t); Bardwell (t); Barnham (t); Blundeston (t); Bungay (t); Burgh (t); Clare (Chilton Street) (t); Cockfield (t); Corton (t); Crowfield (s); Dalham (s); Debenham (t); Drinkstone (p, s); Felixstowe (Walton) (s); Framsden (p); Friston (p); Garboldisham (p); Gazeley (t); Great Thurlow (s); Great Welnetham (t); Herringfleet (s); Holton (p); Ilketshall St. Lawrence (t); Ixworth (t); Kelsale (Carlton) (t); Leiston (Minsmere Level) 3 drainage mills (s); Lound (t); Peasenhall (s); Rattlesden (t); Reydon Quay (Southwold) (t); Saxtead Green (p); Stansfield (t); Stanton (p); Syleham (p); Thelnetham (t); Thorpeness (Aldringham) (p); Walberswick (t); Woodbridge 2 mills (t).

SURREY

Charlwood (s); Chiddingfold (s); Ewhurst (t); Frimley Green (t); Mugswell (x); Ockley (s); Outwood (p); Reigate (Wray Common) (t); Reigate Heath (p); Tadworth (p).

SUSSEX, EAST

Alfriston (t); Chailey (North Common) (s); Cross in Hand (p); Herstmonceux (p); Icklesham (p); Mayfield (Argos Hill) (p); Mark Cross (t); Nutley (p); Patcham (t); Polegate (t); Punnett's Town (s); Rottingdean (s); Rye (s); Stone Cross (t); West Blatchington (s); Winchelsea (p).

SUSSEX, WEST

Angmering (t); Arundel (t); Barnham (t); Clayton ('Jack') (t), ('Jill') (p); Earnley (s); East Wittering (t); Gatwick Manor (t); Halnaker (t); High Salvington (p); Keymer (p); Lowfield Heath (t); Nutbourne (t); Pagham (Nyetimber) (t); Selsey (t); Shipley (s); Washington (s); West Chiltington (s); Weald and Downland Open Air Museum, Singleton—drainage mill moved from Pevensey.

TYNE AND WEAR

Cleadon (t); Fulwell (t); Newcastle (Chimney Mills) (t), (Heaton Park) (t); North Shields (t); West Boldon (t); Whickham (t); Whitburn (t).

WARWICKSHIRE

Chesterton (t); Kenilworth (t); Napton (t); Norton Lindsey (t); Thurlaston (t); Tysoe (t).

WILTSHIRE

Wilton (near Great Bedwyn) (t).

YORKSHIRE, NORTH

Appleton Roebuck (t); Askham Richard (t)—now used as water tower; East Cowick (x); Elvington Brickyard (t)—a pumping mill; Hemingbrough (x); Kellington (t); Kirkbymoorside (t); Osgodby (x); Riccall (t); Skelton (t); South Duffield (t); Stutton (t); Thornton-le-Clay (x); Tollerton (t); Ugthorpe (x); Ulleskelf (t); York (Holgate) (t).

YORKSHIRE, SOUTH

Branton (t); Cantley (t); Carr (Maltby) (t); Conisbrough (t); Fishlake (t); Hatfield (t); Hatfield Woodhouse (t); Moss (Wrancarr) (t); Sykehouse (t); Thorne (t); Wentworth (t).

YORKSHIRE, WEST

Aberford 2 mills (t); Birstall (t); Bramham (t); Darrington (t); Kippax (t); Middlestown (t); Pontefract (t); Swillington (Colton) (t).

BIBLIOGRAPHY

Windmill books are even rarer than windmills. This list includes the important contributions to windmill literature during the past thirty years or so.

BRANGWYN, F. and PRESTON, H. *WINDMILLS*. 1923.

FARRIES, K. G. and MASON, M. T. *THE WINDMILLS OF SURREY AND INNER LONDON*. Charles Skilton. 1966.

FREESE, STANLEY. *WINDMILLS AND MILLWRIGHTING*. David and Charles reprint. 1971.

HEMMING, PETER. *WINDMILLS IN SUSSEX*. 1936.

HOPKINS, R. THURSTON and FREESE, STANLEY. *IN SEARCH OF ENGLISH WINDMILLS*. 1931.

HUGHES, J. *CUMBERLAND WINDMILLS*. Cumberland and Westmorland Archaeological Society. 1972.

LONG, GEORGE. *THE MILLS OF MAN*. 1931.

MAJOR, KENNETH. *THE MILLS OF THE ISLE OF WIGHT*. Charles Skilton. 1970.

SMITH, ARTHUR. *WINDMILLS IN HERTFORDSHIRE*, 1974. *WINDMILLS IN BEDFORDSHIRE*, 1975. *WINDMILLS IN CAMBRIDGESHIRE*, 1975. *WINDMILLS IN BUCKINGHAMSHIRE AND OXFORDSHIRE*, 1976. *WINDMILLS OF SURREY AND GREATER LONDON*, 1976. Available from Stevenage Museum, Herts.

SMITH, DONALD and BATTEN, M. I. *ENGLISH WINDMILLS* (Vol I: Bucks, Essex, Herts, Middx and London; Vol II: Kent, Surrey, Sussex). Architectural Press. 1932.

VINCE, JOHN. *DISCOVERING WATERMILLS*. Shire Publications. 1976.

VINCE, JOHN. *WINDMILLS IN BUCKINGHAMSHIRE AND THE CHILTERNS*. Format Publishing, Ferry Road, Thames Ditton, Surrey (90p post free). 1976.

WAILES, REX. *THE ENGLISH WINDMILL*. 1954.

WAILES, REX. *WINDMILLS IN ENGLAND*. Charles Skilton. 1976.

WATTS, MARTIN. *SOMERSET WINDMILLS*. Agraphicus. 1976.

WEST, JENNY. *THE WINDMILLS OF KENT*. Charles Skilton. 1971.

Printed by C. I. Thomas & Sons (Haverfordwest) Ltd.,
Press Buildings, Merlins Bridge, Haverfordwest, Pembs.